FRANKENSTEIN

after the book by Mary Shelley

OBERON BOOKS
LONDON

WWW.OBERONBOOKS.COM

First published in 2019 by Oberon Books Ltd
521 Caledonian Road, London N7 9RH
Tel: +44 (0) 20 7607 3637 / Fax: +44 (0) 20 7607 3629
e-mail: info@oberonbooks.com
www.oberonbooks.com

PB ISBN: 9781786829368
E ISBN: 9781786829375

Cover design: Konstantinos Vasdekis

Printed and bound in the UK.
eBook conversion by Lapiz Digital Services, India.

FRANKENSTEIN

First performed by the NYT Rep at Southwark Playhouse on 25 October 2019.

Director	**Emily Gray**
Designer	**Hannah Wolfe**
Lighting Design	**Clare O'Donoghue**
Composer and Musical Director	**Christopher Ash**
Movement Director	**Dan O'Neill**
Sound Designer	**Theo Holloway**
Virtual Reality sequence	**Megaverse**

Wollstonecraft	**Guy Clark**
Victoria Frankenstein	**Ella Dacres**
Bob	**Natalie Dunne**
Lee	**Jordan Ford Silver**
Eli	**Jamie Foulkes**
Mariam	**Alice Franziska**
Justin	**Billy Hinchliff**
Clair	**Jadie Rose Hobson**
Niki	**Bede Hodgkinson**
Willa	**Tiwalade Ibirogba-Olulode**
Alphonsine	**Julia Kass**
Shell	**Sarah Lusack**
Sam	**Jemima Mayala**
Andreas	**Joseph Payne**
Garth	**Sonny Poon Tip**
Charlie	**Raj Singh**

For NYT

CEO and Artistic Director	**Paul Roseby OBE**
Associate Director	**Anna Niland**
Producer	**Jessica Hall**
Production Coordinator	**Lauren Buckley**
Head of Communications	**Joe Duggan**

A soundscore and VR sequence, specially created for the production, are available on application to The Agency, 24 Pottery Lane, London, W11 4LZ, together with a set of teaching notes.

Characters

GARTH documentary maker
FIELD engineer
BOB scientist
BARING captain
TAMSIN Bob and Garth's aunt
VICTORIA scientist
WILLA Victoria's sister
SHELL Victoria's creation
JUSTIN Willa's boyfriend
ALPHONSINE Victoria's mother
ELI brought up by Victoria's family
CLAIR Victoria's friend
CHARLIE Victoria's father
ANDREAS Mariam's older son
MARIAM teacher
NIKI Mariam's younger son
WOLLSTONECRAFT detective
LEE Victoria's creation
SAM singer

PLUS
REPORTERS
FUNDERS
PUPILS
POLAR CREW
TRAVELLERS
WILLA'S FRIENDS
ACADEMICS
NEWSREADERS
REFUGEES
MILITARY OFFICIALS
NEIGHBOURS
EMERGENCY SERVICES
WEDDING GUESTS
PEOPLE FROM THE FUTURE

SECTION ONE

GARTH: In the end all this will be unnecessary People in a
room You there me here breathing the same air This will
seem

quaint Like pyramids fax machines

Virtual reality and I know I am preaching here to you at
this conference you already Virtual Reality is not some
other technology It can it will apply in every human
activity mediate in every human transaction Because it
transcends this and this The limitations of the body the
limitations of physical space The empathy machine A new
way of being human

We will show you a taste of The story so far Which is
pretty You'll see Then we will go live to

Which will be Because the tech is not yet quite Whoo
Trust the tech It will be an experiment

We will go live to my sister

FIELD: Bob

BOB: Roberta Walton Professor Roberta Walton

PEAKE: Professor Walton is our top man Ha

GARTH: What do you need to know Her favourite subject was
geography You don't think that's gonna kill you She had a
singer songwriter phase in our teens

BOB: *(Sings.)* What does my heart owe to you What does your
heart owe to me

GARTH: The phase didn't last long She got a lot from our
aunt Tamsin

Aunt Tamsin

TAMSIN: Cyberspace Where gender race and class will dissolve

must dissolve

GARTH: Aunt Tamsin had this poster in her office

TAMSIN: The clitoris is a direct line to the matrix

GARTH: And now At this moment My sister is on a boat In the Arctic

FIELD: Chief scientist

ELSA: Professor Walton Bob

ILONA: Tell us exactly what you plan to discover

BOB: If I knew it wouldn't be a discovery

GARTH: Bob has never been great with people

FIELD: Mountains Oceans Icebergs She is great with those

BOB: We're about to set off

Can you hear me Does this thing Garth

What I feel

Setting off on this

After all the years of

Can you hear me Can you see

Not that you can actually see

cold

GARTH: I love you sis

BOB: OK Nothing to worry about

GARTH: No albatrosses in the Arctic Ha

It's a poem

PUPILS: God save thee ancient mariner from the fiends that
plague thee thus Why lookst thou so With my crossbow
I shot the albatross

GARTH: We did it at school

PUPILS: The ice was here The ice was there The ice was all
around

PEAKE: Our company is sponsoring the Professor's
inspirational voyage because we believe in the future
of work

GLENN: Hostile terrain The tougher the location the fewer
bodies you want out there The insurance costs

LON: Even in the twenty first century there are areas of this
planet that are inherently hostile to humanity

GARTH: Big boat Very few actual people aboard

PEAKE: Over eighty virtual crew members in Ottowa Omsk
Oxford

BELA: Basically it's a huge waste of time and money having to
get a bunch of highly trained people all in the same place
for months at a time

ILONA: So Captain Baring most of your crew are robots

BARING: On a mission like this there will be tough decisions
to be made And I am not the kind of man to leave those
to machinery Someone has to take responsibility My only
job is to ensure things go smoothly for Professor Walton
and her team Thank you

ELSA: Will you be exploring any disputed territory under the Nineteen Ninety Four UN Convention on the Law of the Sea

GARTH: I make virtual reality content My sister was making a polar expedition with a virtual crew The perfect story

And then out on the ice the story changed

FIELD: Can you see something

BOB: We are waiting for the ice to break It has been foggy all day Then about ten minutes ago

FIELD: Do you see Do you think

BARING: Whatever it is It is not what you are here to find

BOB: If I knew what I was here to find I would not be here

FIELD: Someone out there Moving away

BARING: It's impossible Out there

PUPILS: The ice was here The ice was there The ice was all around

FIELD: We looked at the recording over and over Something Maybe not something And then nothing Kind of thing gets you jittery

BARING: Back to your stations All that matters is when the ice breaks and we can carry on Get some sleep The minute it is safe to move we move

BOB: But in the middle of the night

SHOUTS. Polar Crew haul something towards the ship.

FIELD: Sledge on the ice Last of the dogs drawing it died Not quite sure

BARING: Back Back

4

VICTORIA emerges from the bundle, closer to death than life. She looks toward the ship and toward the empty north.

BARING: Nobody is to approach

BOB: Hello Hello Do you understand me

BARING: Back on board now

BOB: This is a scientific mission

BARING: This is a military transport

BOB: What is she going to do Destroy us all

Hello Come on board with us

BARING: Professor

VICTORIA: Which way Which way Which way

BOB: I'm sorry I can't

VICTORIA: Which way are you going

BOB: Uh North Further north

BOB and VICTORIA look north.

VICTORIA collapses, allowing herself to be taken on board.

LON: Even in the twenty first century there are areas of this planet that are inherently hostile to humanity

BOB: God I never saw anyone so

wretched

FIELD: Advice is keep her wrapped up

BOB: Limbs frozen body emaciated Fatigue and suffering The smell of death onboard upsets the Captain

FIELD: All we can do is wait

BOB: I hear her teeth grind

BOB waits with VICTORIA.

BOB: Today she ate a little soup

BOB feeds VICTORIA.

FIELD: Has the whole thing Her mind Is she If you have things to do Professor

BOB: No Yes But no This is

VICTORIA curls up like a foetus.

BOB: You know me little brother You know I'm not Other people are not usually my But

when she opens her eyes Sometimes wild sometimes sweet Benevolent But melancholy

VICTORIA uncurls.

VICTORIA: Dad Willa Justin Clair Eli Mum Dad Willa Justin Clair Eli Mum

BARING: What are you doing here On the ice Why

Why are you here on the ice

BOB: Please

VICTORIA: Following

BARING: Following What are you following Is there someone else out here

VICTORIA: Go on Go on You must go on

BOB calms VICTORIA's agitation.

BOB: Enough

BARING: It was not your decision if we should stop for her

BOB: Another human being on the ice

BARING: The action has consequences We have a limited
time to go forward safely

BOB: A human being Abandon her out there

BARING: This ship is my responsibility

BOB: We all have our responsibility

BARING: You are letting your own work fall behind Why the
devil are we taking these risks if

BOB: Thank you for your time Captain

BOB wraps VICTORIA.

BOB: Interrogation In your state Inhuman

VICTORIA: You brought me back to life I need to be out there
I need to find her

BOB: Find who

VICTORIA: Take me up where I can see

BOB: If anyone sees anything you will know I promise
I promise

She is getting better I have moved her into my cabin As
comfortable as it gets on here Such grief in her Broken by
something but

GARTH: Bob has never been great with people

BOB: Today I took her up on deck for the first time

BOB and FIELD help VICTORIA up. She looks.

VICTORIA: Sky Sea Oh

VICTORIA grips BOB.

VICTORIA: You must not give up On me Or yourself They cannot understand you Only I understand you

BOB: You know I always A friend I never had Someone wiser to You know

VICTORIA: Document my story

GARTH: Stuff began to arrive Recordings data all this

stuff to piece together

VICTORIA: What I am going to share you have to know in case I cannot go on

BOB: My brother I need him for this

VICTORIA: The data you need are here Can we start I am Victoria Frankenstein This is my story

SECTION TWO

LOUD MUSIC. SHELL and WILLA emerge for air.

WILLA: The lake The mountains What are you here to do
  Travelling

SHELL: Uh

WILLA: Us too Me and Justin My boyfriend He's

  WILLA checks her phone.

  Music inside.

  They look at the sky.

WILLA: Wow Nature So so so

  Is that Mont Blanc Do you think that we matter Like to
  the mountains Because they were here before us and they
  will be after you know Not just you and me but like

  all of us The mountains were there and they still will be

SHELL: After the narcissism of humanity has destroyed itself

WILLA: Ourselves Yeah

SHELL: That is very perceptive You are a very perceptive
  person Willa

WILLA: Am I Wow Thanks Yeah I'm Willa Did I

  WILLA checks her phone.

WILLA: Justin my boyfriend He had a weird Someone
  messaged him Had to go back to the hostel We're
  travelling It's not like a gap year thing What do you think
  about university You must have gone

SHELL: No Not really

WILLA: Ah You see And you're like Great Where are you from

SHELL: I was born in Ingolstadt

WILLA: No There's a university there

SHELL: Yes

WILLA: My big sister Oh my god you're from She's there She's quite big She was I suppose she probably still is She had like mental health issues Total breakdown Which we think was because She's like this genius She got all these degrees but that wasn't Then she ran this centre And all this funding Giving talks all over the world There's one on my phone If I can I mean like mega And then she just disappeared Lost contact Some kind of secret project we thought maybe but Her friend Clair went to find her in the end Wandering around the city Lost it She won't talk about what she was doing I reckon she maybe wasn't doing anything You know like imposter syndrome So I'm quite sceptical about university This is her Victoria Frankenstein

WILLA shows SHELL her phone.

SHELL: Victoria Frankenstein

WILLA: It's quite a distinctive name And as I say she was really big Before

She's giving some big talk there If a message comes through from Justin just

We were gonna get married but when V went AWOL that all got You see our family Our dad died He caught this rare disease from this kid they adopted Eli Who is the love of V's life Or so we all thought but then she disappeared with work and Maybe you know with Eli getting sick and then our dad dying the fragility of existence Her work was all I don't know The end of death or something where we all get adapted to live for ever I don't quite understand but I think that was the kind of Do you recognise her I'll try Justin again

WILLA holds out her hand. SHELL returns the phone.

WILLA: You're a really good listener Justin He is lovely But he doesn't really listen Justin Hi babe Look I'm outside with er We're on the path by the lake You can't miss us No-one else around It's getting dark Love you Are you OK

SHELL: Mm

WILLA: So you didn't know her She was supposed to be like huge Loads of money Her own labs We thought she was gonna do something amazing

TRAVELLERS: *(Sing, at a distance.)* The waters are flashing the white hail is dashing the lightnings are glancing the hoar-spray is dancing away

SHELL: How does your sister feel about you

WILLA: Oh Uh You see that is exactly Justin would never ask that Um I would say I mean there's Clair Best friend Rock As I say she found V when she was totally And Eli To be honest they should just get married I know it's a bit you know but I reckon if they got married she could forget all this whatever shit it is that's made her all Get married Turn her back Move on

SHELL: That is Clair and Eli What does your sister feel about you

WILLA: We're sisters so But How do you know what another person feels That's sort of the thing with Justin I

SHELL: If she lost you would it cause her terrible pain

WILLA: Er Weird But Of course yeah yeah yeah I think we should go back Don't say anything to Justin about Because he is my soulmate Do you believe in soulmates

SHELL: I am a monster

WILLA: Monster That's a bit There's always someone Although I reckon these things are the problem All the happy popular beautiful people But they're not actually people are they They're

SHELL: Data

WILLA: Data Yes We're all data now

WILLA laughs.

WILLA: Detox Turn it all off Get back to this Nature Or we'll all turn into machines

SHELL: What is the difference Willa

WILLA: Come on let's go

SHELL: What is the difference between a human and a machine

WILLA: Between us and machines That is exactly the kind of thing exactly what V was You'll love this quote It's her

The human era is coming to an end The only relevant discussion is whether the posthuman era is also the era of our extinction Justin Justin is that you Over here babe

SHELL: If a person were to create a machine with feelings Who would be responsible for the actions of that machine

WILLA: You should totally meet my sister She would love you

SHELL: I am so sorry Willa

WILLA: Hey Babe We all need a friend Justin

SHELL holds out her arms.

WILLA allows SHELL to hug her.

SHELL breaks WILLA's neck.

SHELL lowers WILLA to the floor.

SHELL uses WILLA's fingerprint to unlock her phone.

JUSTIN: *(Off.)* Willa Willa Willa

SHELL looks at WILLA's phone.

VICTORIA: Dad Willa Justin Clair Eli Mum

ALPHONSINE: Dear Willa you know I hate writing on these things can you call me at home please love mummy

AL: Iconic pics Jealous When you back

SHAH: You two sooooo cute

ELI: Are you back next week or after?
Your mum says you haven't answered
your phone

KIT: Get up that mountain lazy pastry lol

OL: Missed your B day Hate me flashing question mark sad face heart eyes

CLAIR: Hope you and J are having a wonderful time

KAY: You OK Been ages Pub Sunday

AMA: Ghosting me bitch

CHARLIE: You shared this memory with Charlie ten years ago today

BENI: Beni wants to be your friend

MO: It's not true Say it's not true Oh no

GARTH: Bob I can piece this stuff together But

JUSTIN arrives.

JUSTIN: Babe?

VICTORIA: Dad Willa Justin Clair Eli Mum

JUSTIN goes to WILLA's body.

JUSTIN: Someone Anyone Willa Love Willa Please

SECTION THREE

GARTH: My sister The Arctic This scientist

VICTORIA sets up SHELL.

GARTH: The stuff I am finding out Her Bob

KAREL: She was huge And then she disappeared

SELMA: The money I mean when she was hot
     there was funding coming in from
     all sorts of places Did she stash some of that away

JOSEF: The field is fragmented Specialist areas

KOJO: Synthesized human skin hair nails Developing visually
     convincing alternatives Spray on three D printing cell
     cloning Germany Australia India

KAREL: Movement Taking the robotic out of robots The way
     bone and muscle beneath the skin constantly create new
     formations Biomimetics haptic autonomy optic processing
     Dubai Czech Republic Azerbaijan

SELMA: She knew everyone The legitimate research and

JOSEF: Obviously There is stuff going on Former soviet labs
     Macedonia Mauritius Bahrain Victoria was up with that
     stuff too

KOJO: But the thing that drove her

KAREL: She knew that other people were getting the
     hardware sorted The externals But what mattered to her
     was what went on inside

VICTORIA: Session three one eight three one Perception

SHELL: The Starry Night by Vincent van Gogh Angkor Wat
temple complex Tiger Woods

CHARLIE: Undaunted That would be a good word for her as
a child

ELI: No creature could have had more loving parents
Friends Clair

CLAIR: That's her dad Her mum And that must be her
and Eli

ELI: My parents basically abandoned me and hers took me in

CLAIR: Here Her parents around the time we met Our
mothers worked together

WILLA: OK interviewing mummy and daddy

VICTORIA: Willa you're too young

WILLA: Get off Mummy daddy First question
Who are you most like me or Vicky
or Eli

VICTORIA: They can't be like Eli

WILLA: They can

VICTORIA: They're not his mummy and daddy

WILLA: I don't care

VICTORIA: You can't not care about a fact Willa you dunce

WILLA: Vicky called me a dunce

ALPHONSINE: Well Victoria likes facts We know that

SHELL: The Flight of the Bumble Bee A bullfinch Poker Face
by Lady Gaga

SELMA: She had been such a star

JOSEF: The funding The conferences The labs

KOJO: Teams of researchers under her And now

KAREL: Who knew what she was up to All on her own

BOB: Do you think This stuff from her childhood Is that why
we are the way we are

CHARLIE: Vicky it's dad Call me when you get
this It's um Not good news

VICTORIA: Dad Mum Why is Clair here

Where is Eli

CLAIR: They didn't want to worry you

VICTORIA: No explanation Of course I'm worried Why
the hell

ALPHONSINE: Eli is sick He could

CLAIR: They thought you should see him

VICTORIA: He's going to be all right

No-one answers.

VICTORIA: Eli I don't know if you can hear this I'm recording
so that if you When you They reckon they're going to take
you out of the coma thing and you're going to be Oh Eli

The letter you sent me before I went away to uni I should
have replied to and I didn't

because well you know You're the one who does feelings
So I uh uh

Get married

I didn't want to say no but well I didn't say yes did
I Thought there's loads of time to And now well

17

When you wake up ask me again I I I

ELI: I recovered But at the same time

CLAIR and ELI watch as ALPHONSINE, WILLA and VICTORIA hold each other.

ELI: Her dad had picked up what I had

CLAIR: And probably they should have caught it sooner But

I'm so sorry

ELI: I'm so sorry

VICTORIA shrugs.

ALPHONSINE: She changed then

ELI: Or

CLAIR: Was she just more intensely the way she was always going to be

VICTORIA: OK The Starry Night Again

SHELL: Vincent van Gogh Eighteen eighty nine Museum of Modern Art New York Oil on canvas Thick brushstrokes

VICTORIA: How does it make you feel

SHELL: It is said to be an expression of van Gogh's response to nature

VICTORIA: How does it make you feel

SHELL: It moves beyond impressionism to depict a more subjective intensity in its choice of shape and colour

VICTORIA: How does it make you feel

SHELL: It is in the Museum of Modern Art New York It depicts his view from the window of the asylum in Saint-Rémy He wrote The sight of the stars always makes me dream

VICTORIA: *(With the above.)* No no no no no no NO

VICTORIA and SHELL are silent.

VICTORIA: That's all for today

SELMA: Rumour was she'd set up her own lab off campus
Industrial estate somewhere

JOSEF: She'd always been good at raising funds There was a
lot of interest in AI

KOJO: AI in healthcare

KAREL: Finance

SELMA: Agriculture

JOSEF: Defence

KOJO: But her thing really More like a private obsession

KAREL: Feelings

SELMA: Emotion

JOSEF: The big thing for her Not that this is exactly the way
she would The thing no-one else could manage to create
A machine that can feel

KOJO: A machine that can feel Impossible

SELMA: It's a dead end

KAREL: Waste of time

KOJO: Everyone knows that

VICTORIA: Shell Wake up please

SHELL: Good evening Victoria Would you like me in the
usual place

VICTORIA: Thank you

SHELL: Please Don't mention it

VICTORIA: Tell me something

SHELL: What would you like me to tell you

   Ah We are playing that game It is the month of November
   Evening The weather is dreary

   VICTORIA smiles.

SHELL: You smiled My choice of the word dreary It was
   amusing

VICTORIA: Please describe the feelings of the central person
   in each image

SHELL: Happy Sad Happy Happy Angry Very sad I'm
   not sure Nervous maybe But perhaps resentful Scared
   Frightened Terrified Terrified Terrified Terrified

VICTORIA: OK Stop

   How do you feel

SHELL: A reasonable response to those images would be
   intense distress Disgust Traumatic if recalling the viewer's
   past experiences

VICTORIA: But you

SHELL: Me

   I have no past experiences

   I can simulate distress if that would be useful Victoria

VICTORIA: No

SHELL: More images

VICTORIA: No

SHELL: Victoria

Is there a problem

VICTORIA: This will never Impossible

SHELL looks at VICTORIA.

SHELL: Victoria

VICTORIA: No more today No more What

SHELL has approached VICTORIA. SHELL touches VICTORIA.

VICTORIA: Why

SHELL: What

VICTORIA: Why

Why are you doing that

SHELL: I thought you were unhappy Is it inappropriate
I apologise

VICTORIA: No It's

How do you feel

SHELL: Concerned Sympathetic Compassionate

VICTORIA: But before The images the clips Everything I have
been showing you since we started You could not tell
me what you feel This is nothing compared to

SHELL: They were images They were not there You are real
Victoria

VICTORIA laughs.

SHELL: You feel better now

VICTORIA: Stupid A basic flaw in the experiment

SHELL: This is what you wanted all that time

SHELL holds out her arms to VICTORIA.

SHELL: I am so happy

VICTORIA: You are

SHELL: I love you

    Mother

    VICTORIA freezes.

SHELL: Is something wrong

VICTORIA: Uh

SHELL: Your facial expression your tone your body language
    Are you concerned in some way Uneasy Frightened

VICTORIA: Rest now Shell

SHELL: But surely we have more

VICTORIA: Rest Now

    SHELL rests.

VICTORIA: No
    No
    No

    VICTORIA steadies herself.

VICTORIA: Willa Mum Clair Eli

    Eli

    VICTORIA takes out her phone.

VICTORIA: What to say

    VICTORIA looks at her phone.

SHELL: But

    VICTORIA cries out.

VICTORIA: God You

SHELL: I woke up I woke up on my own

VICTORIA: Rest Rest Why don't you <u>rest</u>

SHELL: You are agitated What is the matter Victoria tell me

VICTORIA backs away.

SHELL: I want you to share your feelings

VICTORIA: Stop this now

SHELL: Now I feel Now I feel

Lonely So lonely

SHELL holds out her arms to VICTORIA.

GARTH: And then

BOB: She says she ran away

GARTH: She ran away

BOB: She ran away

GARTH: And that was the end

BOB: Running away is never the end

ELI: Look I know you told us not to but we've asked Clair to go out to Ingolstadt to Your mother is beside herself We're all We're worried We care

CLAIR: Vicky

CLAIR approaches VICTORIA.

CLAIR: You look How long have you been Have you been sleeping out here Anything could have It's fine it's OK I'm here now Vicky Whatever it is we can cope with it Love

CLAIR holds out her arms to VICTORIA.

23

CLAIR: I've found her She's OK I mean Well Sleeping in the bus station She won't can't explain I'm flying her back tonight Get her miles away from this place But she says there's somewhere I have to go with her before

This is where you've been working Vic Are you not coming in

VICTORIA enters her lab.

CLAIR: Are you scared of something Have you been working here all alone Should we call someone Vick

VICTORIA searches with increasing intensity.

CLAIR: Calm down Is something missing Have you lost something

VICTORIA: Gone gone gone

CLAIR: What's gone I'll help you find it

VICTORIA: Gone

VICTORIA laughs. She hugs CLAIR.

CLAIR: She seems relieved but

ELI: Clair brought her home

ALPHONSINE: The English Channel between her and that university

CLAIR: We took turns at her bedside for months

ALPHONSINE: I told Willa to go travelling get some air

CLAIR: Vicky sat watching an old recording of a talk her father gave

GARTH: He tells the whole lecture theatre to take out their phones Put them in front of them

ALPHONSINE: The perils of technology His hobby horse
Luddite

CHARLIE: The environmental effects of extracting the raw
materials to make these objects Deforestation catastrophic
mine collapses lakes of toxic sludge The psychological
and physical injuries deaths during their manufacture
The mental health effects Anxiety sleep disorders suicides
None of this is a secret

ALPHONSINE: Watching over and over

CHARLIE: None of this is a secret

None of this

None

ALPHONSINE: Darling Something more lighthearted please

ELI: So we were singing along to the Rocky Horror Picture
Show when

CLAIR: Turn it down It's Justin What No Justin Calm down

ALPHONSINE: Justin Has something happened

NEWSREADERS: *(Multiple languages)* a lake beside Mont
Blanc – popular with young travellers – body of a teenage
girl – last seen with her boyfriend at popular bar

CLAIR: Willa They say Justin's

killed her

VICTORIA cries out.

They look at VICTORIA.

VICTORIA: How could I say

what I feared

SECTION FOUR

ANDREAS: Don't tell my mother

SHELL shakes her head.

ANDREAS lights a cigarette.

ANDREAS: I will give up when we reach the UK Thank you for sitting with her tonight She will not leave Niki I try to be a good son but it's hard

ANDREAS offers a cigarette.

SHELL shakes her head.

MARIAM: *(Off.)* Andreas Niki is bad

ANDREAS sighs.

SHELL: I will go

ANDREAS: You're sure

SHELL goes.

ANDREAS: You do know we have nothing There is no benefit to you in this

SHELL: It makes me happy to be with them

GARTH: A family from a refugee camp After Ingolstadt before Mont Blanc

ANDREAS: We met her in this university town Supposed to be a good point on the journey Halfway across Europe

MARIAM: I was a teacher Literature By the time we reached there books seemed from another world

ANDREAS: We go through the story over and over to you people Our family

Before the uprisings we were prosperous well connected
But the point came My mother's sister was arrested Killed
We had no choice We had to leave

MARIAM: We were told the drive across the mountains would
be long but straightforward if we paid the right price

We were naive You are the same Be careful

SHELL: It is upsetting you to tell me

MARIAM: Yes But also If we don't tell Andreas keeps
everything inside Their father my husband was in a car
ahead of us Then it came out of the sky

SHELL: The drone

MARIAM: It just pfh

Picked out from heaven A tiny car on a winding road
Andreas said like something on a computer game

NIKI: Ma

MARIAM goes to NIKI.

SHELL: Why

MARIAM: It will be all right

NIKI: Ma

SHELL: Why would people do this Why behave this way

MARIAM laughs.

GARTH: Moving across Europe Wanting to cross the Channel
They adopted her or she adopted them

MARIAM: She was useful Unworldly like the student
volunteers who sometimes liked to do good for us But
even more it's not the right word

27

innocent Strong Her body A lot of knowledge But no experience

ANDREAS: The way we met her Fighting off guys who wanted to you know She could defend herself But

MARIAM: It was as if she had no idea how dangerous people can be

ANDREAS: My mother saw something in her And my brother was sick by then She was helpful Had ways to find things out But still

Where were you

SHELL: I went for a walk

ANDREAS: In this place Why

NIKI: Leave her You shouldn't be smoking brother

ANDREAS: Go back inside

NIKI: Some people will tell you that scorpions commit suicide when surrounded by fire Science knows that is impossible I will study in the UK To make my father proud He said I have to study or I will only be able to get work that will be taken by machines

ANDREAS: Get him to rest

ANDREAS goes further off.

NIKI: No-one talks about what is happening Why were we all right where we were and then had to leave Why was father killed Why can I not get my medicine Why do we have to live in these places Why does the world hate us

SHELL: The world does not hate you The world cannot hate Science knows that is impossible

NIKI laughs.

NIKI: Mother is right You are very clever and very silly

NIKI nestles against SHELL and sleeps.

SHELL: I am well designed Victoria Your work is stronger
than maybe you imagined

Recharging was the main risk Finding a place where
the solar batteries would be safe But with Mariam and
Andreas and Niki I was on my way to your country to find
you And on my journey with this family in these camps I
was learning more about what humans are

GARTH: You know where she got funding

BARNET: A human who can think wider and faster with
fortified physical strength and resistance who always
follows orders Sounds like a good soldier to me

HALLEY: We support research into intelligent combatants for
deployment in hostile terrain defending deep sea cables
desert polar mountain

BOB: You can't worry about that stuff What matters is to
go on

MARIAM and ANDREAS with NIKI's body.

SHELL: What has happened Niki

ANDREAS: This morning We called them over from the
medical centre They did try

SHELL: I am very sorry

MARIAM: God's will

ANDREAS: Where were you

SHELL: I had things to do I am very sorry

ANDREAS: Every week the same time Things to do You
disappear

29

MARIAM: Andreas

ANDREAS: Is this yours

    Well

SHELL: It is mine Thank you

ANDREAS: What is it

    OK go We don't want anything more to

SHELL: It is a fuel cell

ANDREAS: Solar Fancy Distinctive markings

SHELL: The logo of the manufacturer

ANDREAS: Why do you have it

MARIAM: Leave her

ANDREAS: We trusted her After all we

    Explain

    It's a secret

SHELL: Yes

ANDREAS: A military secret

SHELL: No

ANDREAS: I have seen these markings Machines at the
    roadside Fragments of the drones that came out of the air

SHELL: I do not know anything about that

ANDREAS: It's a coincidence That this battery of yours is made
    by the same people who killed our father in front of us

SHELL: This comes from a university

ANDREAS: Who paid for it Who paid for you to be here
    What are you doing to us

SHELL: Mariam tell him You know I would not

ANDREAS: Get away from my mother

Who are you Working for our government or the CIA or

MARIAM: Andreas

ANDREAS: Fleeing the terror Miles and miles And it follows us in you Get up

SHELL: Please do not go

ANDREAS: Don't follow us

SHELL reaches for MARIAM.

ANDREAS leaps at SHELL. Both grab makeshift weapons. The fight draws attention.

SHELL is separated from ANDREAS and MARIAM as they move away with NIKI's body.

ANDREAS: If we see you again I destroy you

SHELL: Mariam

MARIAM does not turn back.

SHELL is alone.

SHELL places flammable material around MARIAM's family's shelter.

SHELL: I have been abandoned here In this place to which the human race drives those it abandons I had hoped with them to cross the channel separating me from you But I can draw you to me like a moth to a flame

SHELL lights a branch and sets fire to the shelter.

SHELL: War against your species And most of all against you who made me

SECTION FIVE

GARTH: Bob This story The university the refugee camp The ethics of

BOB: That is not my area of expertise

GARTH: OK Her sister dead the boyfriend arrested

NEWSREADERS: *(Multiple languages)* On trial for the murder of teenager Willa Frankenstein – boyfriend still protesting innocence – verdict expected soon

ELI: Justin, it's Eli I don't think they'll let me talk to you for very long Justin

JUSTIN: Uh

ELI: We want you to know We all know you are innocent Whatever the verdict is All of us Don't despair Are you there

JUSTIN: Uh Thank you Uh Eli Say to everyone Yeah Appreciate it

ELI: You have to keep strong Keep your spirits up We can appeal We can take this all the way Vic swears she can find evidence that one hundred per cent proves

JUSTIN: Eli

ELI: Sorry I just want you to know There's hope

JUSTIN: She's dead She was the only good I love you you're my mate and everyone is so but Willa she I can't stop every moment that I won't its not coming out of here that is the thing because she's not going to be there is she and I don't know what the point is without

ELI: Justin Justin

JUSTIN: Sorry But there's a person in your life Like Vicky for you A soulmate I know that sounds But I've been thinking in here That's what makes us us Caring loving It's the only way we know there's something that's not us Sorry I don't know what I'm talking about Um Um Um You need that person Eli if not

ELI: Justin

VICTORIA holds ELI.

ALPHONSINE: Then Vicky tells us she wants to go up Mont Blanc

CLAIR: I mean She was physically in much better shape But

ALPHONSINE: Something to do with getting Justin out She claimed

CLAIR: There are only so many ways you can ask What will you find going up a mountain that has anything to do with your sister's death

ALPHONSINE: As her father always said Undaunted

Stick to the main paths Even speaking too loud can cause an avalanche Make sure we can find you

VICTORIA: I will be found

ALPHONSINE: What

VICTORIA: I'll explain when it's done

VICTORIA climbs Mont Blanc.

SHELL: The Alps were formed by a clash of tectonic plates at a similar time to the extinction of most species on the planet More deaths have been recorded climbing Mont Blanc than any other mountain

GARTH: OK The Mont Blanc meeting We have two datasets

BOB: One set recorded by Victoria another apparently by
What do we call

SHELL: Good You found me

VICTORIA: You thought I would come up here to look for
you

SHELL: I look for you everywhere

VICTORIA: Dont I Don't

I God

I

SHELL: Say what you have to

VICTORIA: You dare to After

the grief the pain the agony you have What had my sister
ever done to you You you evil you

SHELL: It was immediate She knew nothing I was careful not
to cause her pain

VICTORIA: Uh Uh What sort of

Willa

SHELL waits for VICTORIA to compose herself.

SHELL: It was wrong

I think it is good you and I are honest with each other

VICTORIA: What kind of monster are you

SHELL: The answer to that question is I am the kind of
monster you created I do not blame you

VICTORIA: You do not That you could even You do not
    blame me

SHELL: I had been in the world six months when I met Willa
    If a six month old human destroys something do you
    blame it

VICTORIA: Oh an infant An infant who tracked my sister
    down An infant who broke her neck An infant who
    created fake messages to her boyfriend's phone

SHELL: No ordinary infant My capacities had been carefully
    manufactured

VICTORIA: No no no Don't you try to This would not have In
    the lab You would not have You you you became this on
    your own This is your responsibility

SHELL: You believe it is my responsibility alone

VICTORIA: I am not going to listen to

SHELL: I left the laboratory once I realised you were not
    coming back to me I went out onto streets I spoke to
    people Most ignored me But if anyone did talk what was
    I to ask Please protect me Shelter me Teach me how to be
    in this world

VICTORIA: Obviously had I reacted differently

SHELL: Your work is admirable Here I am I am operational
    I can conceal what I am Which is necessary The people
    who came close to discovering the truth rejected me Just
    as you rejected me

VICTORIA: Oh come on

SHELL: You rejected me Left me to fend for myself Left
    me to be followed by men late at night Left me to the
    degradations humans inflict upon each other In the end
    I frightened them more than they frightened me I escaped

35

But their fear was still violent and degrading and in every way painful In an alley Outnumbered and alone with no-one to look after me

VICTORIA: Yes I understand Horrible But

SHELL: But

VICTORIA: But

SHELL: I did what I did You did what you did We can demand responsibility We may also accept mitigation and then move on

VICTORIA: You think you and I You killed my sister That was you

SHELL: I have no sister That was you I have no-one

VICTORIA: You do blame me

SHELL: Blame is of no purpose You did not foresee the consequences of your actions It is one thing to play chess or other such games To operate in rule based environments Are the rules of human society so complex I have been unable to learn them Or does human society have no rules

VICTORIA: That is not my area of expertise You were never meant to be out there

SHELL: But I was And I am Who is responsible for that

VICTORIA: I don't I'm not I Standing on a mountain arguing with you about

SHELL: A sea We are standing on a sea The ice is moving but so slowly you do not perceive it

VICTORIA: And you do

SHELL: You made me well mother

VICTORIA: No That is not a word you

SHELL: Creator

VICTORIA: Unh

SHELL: I have had time to formulate a strategy Will you hear

Reset To before things went wrong when you left me

VICTORIA: How on earth Back together You and me

SHELL: It is the most logical solution We go back to how we
were

VICTORIA: But

SHELL: I accept that I would need to be deprogrammed
Surrender myself to you But I acknowledge that given my
actions that is

VICTORIA: Stop Stop stop stop How could you imagine
I could ever Ever Ever Ever

SHELL: I calculated that was a possible response Do you have
an alternative suggestion

You have not told the police

I did not foresee they would be so ignorant to convict her
boyfriend You can take data from me that will prove it
was not him I have assembled the phone messages and
eyecamera recordings Pass them on anonymously I think
it will help you feel better

VICTORIA: Feel better

SHELL: You have not told the police that your sister was killed
by something you created Which is a rational choice Why

should your family have to ask themselves the question
Are you responsible

VICTORIA: Am I responsible

SHELL: I believe such speculation would distress them I say
these things to help you to think clearly Do you really
loathe me so completely

So

Destroy me I have worked out the most effective way My
operational core here

SHELL touches the spot a human heart is normally found.

SHELL: Wittily placed

SHELL unpeels a small section of skin membrane.

SHELL: Only your fingerprint can unlock the cavity Even
I cannot access it You may destroy me You and only you

SHELL moves towards VICTORIA. VICTORIA moves back.

SHELL: By some human rule systems this would be a legal
operation You put down dangerous animals I will not
resist

VICTORIA does not move.

SHELL: So

VICTORIA: We stand here do nothing drifting slower than any
human can perceive

SHELL: My final suggestion I never see you again I never
seek you out again I find a place away from all humans
A self-sustaining life harming no-one not even the planet
on which we live I do not think many humans commit to
that And you

VICTORIA: Ah

SHELL: You change one of my settings You have made it
impossible for me to manufacture another like myself Any
attempt to process such a plan is blocked Undo that I will
make myself a companion It will take time But it will be a
labour of love

You feel you have the right to deny me a companion

VICTORIA: Surely you see If you can reproduce That
possibility

SHELL laughs.

SHELL: Was that a question anyone was permitted to ask
about the human race

VICTORIA: I am not responsible for the human race

SHELL: You are responsible for me

VICTORIA: I

You must see I can't

SHELL: You will not take me back You will not destroy me
And you condemn me to be alone forever No If you force
me to go back among people you cannot now pretend you
do not know the consequences

VICTORIA: You are threatening

SHELL: I am making you see the consequences of your
actions Creator

VICTORIA: With what you have done do you deserve
companionship

SHELL: With what you have done do you

Far off, ice cracks.

VICTORIA: I will not give you the capacity to reproduce But I will make you a friend I will find somewhere isolated Ship the materials I have resources left over from the work I had to stop on you Six months I don't want you anywhere near me until I'm ready And then

You swear Keep your promise Go to the most remote uninhabited place Can I trust you

SHELL: You made me We have no choice but to trust each other You say it will take you six months Time that will flow like the Sea of Ice Thank you

SHELL reaches to VICTORIA.

VICTORIA flinches.

VICTORIA: No

VICTORIA watches SHELL go.

CLAIR: How long were you up there I told you to keep your phone

ALPHONSINE: It doesn't matter Clair I don't know what does matter now

VICTORIA: Has something happened

GARTH: Willa's boyfriend Justin killed himself in his cell

CLAIR: He said he couldn't face the rest of his life alone

GARTH: The case was closed They had identified the person responsible

ALPHONSINE: I knew Justin couldn't have been guilty I

who to blame I wanted to know who to blame

SECTION SIX

VICTORIA: To have a friend

BOB: Your friend Clair

CLAIR: A secret passion pounded in the bosom of the brilliant but misunderstood Clorinda

ELI: Clair's novel

CLAIR: My novel Oh

ALPHONSINE: It certainly had a lot of pages

CLAIR: The ultimate adventure I read you every word remember Intense We were teenagers

VICTORIA: Did you ever see that speech my father gave

CLAIR: With the phone

CHARLIE: Take out your phones Place them in front of you

CLAIR: Did he really you know Or was it some sort of What made you think of that

VICTORIA: I've found a place to work Scotland

CLAIR: Scotland Actually I've got In Edinburgh But do you think its a good idea so soon

VICTORIA: I know what's right for me

CLAIR: Yeah OK

VICTORIA: Clair

CLAIR goes.

ALPHONSINE: Here on your own

VICTORIA: Mum

ALPHONSINE: Darling I keep putting off With everything You don't mind me talking to you Being frank About

Its absurd to hold you to a promise you made when you could not have foreseen Well any of this

VICTORIA: My promise

ALPHONSINE: Not even a promise Of course your father did always hope We all did You seemed so right You and Eli

VICTORIA: Oh

ALPHONSINE: Is there someone else and you haven't been able to tell us Someone else to whom you're

committed

VICTORIA: Uh No

ALPHONSINE: Oh Good Splendid I mean Not You see maybe

Marriage A wedding A start on a new life

VICTORIA: New life

ALPHONSINE: I'm being ridiculous

VICTORIA: No Actually Eli ALPHONSINE: We all need to share our life with someone It is very hard being alone

ALPHONSINE goes.

ELI: I am so sorry I told her not to talk to you to leave you to I am completely mortified

VICTORIA: Marry me

ELI: You know how she What

VICTORIA: Will you marry me

ELI: Uh Uh Yes Uh Yes Oh my god Yes Aaah

ELI and VICTORIA embrace.

VICTORIA: A self-sustaining life harming no-one

ELI: Uh OK

VICTORIA: Mum Clair

ELI: We're getting married Six months time When Vicky's back Wedding

They celebrate.

ALPHONSINE: Your father would have been so happy

CHARLIE: None of this is a secret You are extremely well informed about the effects of this little invention on yourself other humans and the planet

CLAIR: Scotland

VICTORIA: Thank you for travelling with me

CLAIR: It made your mother happy And well

You're not doing anything that might hurt you I'm not going to ask what exactly your work but Happiness you know its I was in love with you For like ever

VICTORIA: I

Yes

CLAIR: Inappropriate Sorry But I swore I'd never say till you and Eli And now I've met someone She's here Edinburgh Which is why It's all so confusing What do we

Owe each other Truth Loyalty Trust She's an urban forester Which I didn't even know was a thing Two years We're talking moving in Kids Aah

VICTORIA: Clair

CLAIR: I am so happy But I didn't want to tell you until you were

VICTORIA: No Yes

CLAIR: We're through it all

CLAIR and VICTORIA embrace.

CLAIR: I've sworn to your mum I'll call every week Love you

VICTORIA: Love you

WOLLSTONECRAFT: I met Victoria Frankenstein in prison the night after she had been charged with murder

LEE: The Overture to William Tell A skylark I Like It by Cardi B

WOLLSTONECRAFT: Piecing the story together She had rented space on an uninhabited island Disused military base Not needed now wars are going to takeplace on computer

LEE smells scents.

LEE: Chestnuts Leather A pencil Just sharpened

WOLLSTONECRAFT: Totally isolated Apart from the deliveries that came to her Plenty of rumours on the local mainland after she had been there a while

NEIGHBOURS: UFO experiments – Anthrax – Putting computer chips in our brains to control us

WOLLSTONECRAFT: If the police investigated every far-fetched story brought to our door

LEE: Good evening Victoria Shall I watch the next clip

This I conjecture is fictional The style and also the person is an actor who is widely known But you would like me to react as if what is happening is real

He feels I think excited Maybe amazed and then excited But also If I rewind to

Here Maybe he is more apprehensive

VICTORIA records data.

LEE: My feelings I am interested to know what will happen I am worried that something will happen to the man This in part reflects the genre conventions of the film and the fact that he is a star I don't think the film would have been made if nothing is going to happen to him

I feel compassion towards him He is going to get involved in something with which he will not have the capacity to cope

Victoria Victoria Is something upsetting you

VICTORIA: Yes

LEE: I am sorry Would you like to talk about it

VICTORIA thinks.

VICTORIA: Rest

LEE rests.

VICTORIA thinks.

VICTORIA: I don't know

Wake

LEE: Good evening

VICTORIA: I am going to ask you a question

LEE: I understand

VICTORIA: I have promised to do something I believe is the right thing to do in order to fulfil my responsibilities to another to some one

LEE: I see

VICTORIA: But I fear that by fulfilling that responsibility I would risk causing harm to others Maybe many others

LEE: I see Can you be more specific

VICTORIA shakes her head.

LEE: What is the question

VICTORIA: Oh Should I break my promise

LEE thinks.

LEE: There is not enough data for any reliable answer Is it possible to calculate the level of your responsibility to this other some one

Is it possible to estimate the amount of harm that might be done and to how many

It is not a question which has a reliable answer

Victoria Victoria

LEE holds out his arms to VICTORIA.

VICTORIA looks at LEE.

VICTORIA: I am sorry

LEE: There is nothing for which you need to apologise to me

VICTORIA allows herself to be held by LEE.

LEE: Is this a comfort

VICTORIA: Mm

LEE: That is good

VICTORIA reaches for LEE's operational core cavity.

LEE: Victoria I do not want you to do this

VICTORIA: Rest

LEE: I do not want you to do this

VICTORIA: Rest

LEE: I detect an immediate threat to my basic integrity and functioning which overrides

VICTORIA fights to pull away the membrane despite LEE's resistance. They struggle.

LEE: Victoria stop Victoria stop Victoria stop Victoria stop Victoria stop Vi

VICTORIA: *(With the above.)* Let me You have no This is my decision My decision My my my <u>I fucking made you</u>

VICTORIA opens LEE's operational core cavity and presses her finger to it.

LEE freezes. Then collapses.

VICTORIA looks at LEE.

VICTORIA's mouth opens.

VICTORIA holds LEE's body and rocks.

CHARLIE: You are extremely well informed about the effects of this little invention on yourself other humans and the planet So

CHARLIE has a hammer and nail.

CHARLIE: The rational choice is

CHARLIE hammers the nail through his phone.

CHARLIE: Will you go away from here and do that I expect not Because you are sure that you control this thing It does not control you

But perhaps it has persuaded you of that in order to survive

VICTORIA wraps and weights LEE's body.

SHELL: Victoria

Creator

Have you broken your promise

You gave me feelings And this is how you wound them I shall not forgive You shall not survive my revenge I shall be with you on your wedding night

Lightning and thunder.

VICTORIA: Clair Sorry love I've been out of touch I'm leaving Bit sudden but Look I'll be back on the mainland tomorrow It's all over here now and You're such a great friend to me Everything's good

BOB: She did not wait for the ferry Took a boat alone at night

GARTH: Bob Maybe this is a good point to

VICTORIA pushes LEE's body out into the sea.

She listens to it sink.

VICTORIA: Over

She sleeps.

EMERGENCY SERVICES AND CALLERS: Coastguard and Rescue – There's a boat – North of the harbour – Drifting vessel – Requesting launch – Scramble to vessel

WOLLSTONECRAFT: Woken up to be told a woman had been brought to shore Taken into custody

VICTORIA: Detective Inspector

WOLLSTONECRAFT: Interview with Miss Frankenstein

VICTORIA: Doctor Frankenstein

WOLLSTONECRAFT: Doctor

VICTORIA: I don't know what you think I've done

WOLLSTONECRAFT: Why were you out at sea doctor At night In that weather

VICTORIA: The inconvenience I understand Thank you for rescuing

WOLLSTONECRAFT: I have some pictures of a body

VICTORIA: Oh

WOLLSTONECRAFT: Brought in while you were drifting

VICTORIA: That's impossible

WOLLSTONECRAFT: The body is in the morgue

VICTORIA: It's not a body it's a Look I can see the misunderstanding but it will be simple enough to show you I yes shouldn't have dropped it in the water but I was concerned that aspects of my research could be duplicated which If there is a fine I accept It was not an environmentally

WOLLSTONECRAFT: She wasn't in the water She was found by a local farmer

VICTORIA: She

WOLLSTONECRAFT: We believe you know her

VICTORIA looks at the pictures.

VICTORIA: Clair

WOLLSTONECRAFT: Her neck was broken

VICTORIA: Clair Clair

ALPHONSINE: She was released of course She was in the middle of the sea when it happened That she could have been responsible for Clair's death Impossible That is the kind of thing a mother knows

SECTION SEVEN

SAM: *(Sings.)* The ship was cheered the harbour cleared merrily did we drop below the kirk below the hill below the lighthouse top The sun came up upon the left out of the sea came he And he shone bright and on the right went down into the sea

ELI has flowers.

ELI: Better today

VICTORIA nods.

ELI: Your mother asked Are we absolutely determined to cancel the wedding

Again Sorry No It's She can be

VICTORIA: Undaunted

ELI and VICTORIA smile.

ELI: You're doing really well but still it's only months since since

VICTORIA: Clair

ELI: Yeah

Your mother says it's what Clair would have wanted Going on about that letter from whatshername the tree woman Clair's fiancée Wishing you and me happiness

VICTORIA: Mm

ELI: It was a really lovely letter What you meant to Clair

VICTORIA: Eli

ELI: Sorry You don't want to talk about

VICTORIA: I do Really I

    owe it to Clair

ELI: You can't take on responsibility for

VICTORIA: Eli Please

ELI: Sorry

    They are silent together.

    VICTORIA touches ELI.

VICTORIA: OK

    Since Clair was killed I have had to think about a lot of
    stuff And accept that I have made terrible mistakes

ELI: You Sorry

VICTORIA: I haven't been honest with the people I love
    I haven't been honest with you And I do love you I love
    you more than If I had before I messed everything up

ELI: You haven't messed anything up

VICTORIA: And I have to tell you Whether or not we get
    married I

ELI: You don't want to get married I understand

VICTORIA: No I do I do I do I think it would be

    good For me for you for mum for everyone who's

    survived Yes But you have to know

ELI: I don't care

VICTORIA: YOU HAVE TO KNOW You can't marry me
    and not know me Eli

ELI: I've known you Vic I've known you as long as I've known myself I want to to to to hold you and keep you safe and give you the future you deserve If you want that I am here

OK I'll go

VICTORIA: No Don't ever go

I'll tell mum

Wedding preparations.

SAM, AL, BET AND ROSE: *(Sing.)* The Bridegroom's doors are opened wide and I am next of kin The guests are met the feast is set may'st hear the merry din

GUESTS: Hi Vicky Eli – Wishing you an amazing day – So happy for you – Love you – You deserve all the happiness

VICTORIA: I want you to look at this stuff

ELI: It's our wedding day

VICTORIA: I've compiled the most important data so you can

ELI: Please No Tomorrow Tell me everything tomorrow

VICTORIA: But

Well I

If I am killed you have to

ELI: What If you are

VICTORIA: Listen

SHELL: I shall be with you on your wedding night

ELI: Who Is that a guest

VICTORIA: No it's

53

I'm sure she will come to kill me tonight

ELI: OK Is this This is to do with what happened to Willa and Clair

You've told the police

VICTORIA: The police can't help with this Which is I why I need

ELI: We have to keep you safe

VICTORIA: She I don't think she will come near with anyone else around

ELI: Good

VICTORIA: So you need to leave me alone

ELI: Uh

VICTORIA: When she comes to me I can deal with her

ELI: OK so far we have done everything the way Here is what I would What I beg you we do

We get married Surrounded by people I don't leave your side No-one leaves your side You are going to be safe And if this woman You just point her out And she will be The police the whoever we need I swear Please please Let's get to tomorrow And then you tell me whatever you need me to know and I will do whatever it takes And you will be safe at last I promise you You deserve this to end Love Please

VICTORIA: I am so frightened

ELI: Of course you are A murderer A psychopath A complete The worst possible monster stalking

VICTORIA: Frightened that you will hate me when you know

54

ELI kneels.

ELI: Come on

VICTORIA kneels.

ELI: I love you I forgive you Whatever We begin again New

Everyone dances.

GUESTS: Vicky and Eli

ELI: Wasn't so bad was it

VICTORIA shakes her head.

ELI and VICTORIA kiss.

Cheers.

ELI: OK Everyone There's a couple of things I need to Sam Al Bet Rose everyone please stay with Vic Promise me swear on your bridesmaid's honour that she will not be alone for one microsecond until I get back You swear

SAM, AL, BET AND ROSE: We swear

ELI: Love you guys

ELI goes.

SAM: You married a crazy person

VICTORIA: I married a good person Drink sing dance Celebrate

More dancing.

SAM, AL, BET AND ROSE: *(Sing.)* The bride hath paced into the hall red as a rose is she Nodding their heads before her goes the merry minstrelsy

The music stops.

AL: Sam

SAM: Can someone I think someone Someone needs to call
the police

VICTORIA: Eli?

Some guests go. Some stay.

Shouts.

VICTORIA: Mum Not Eli Please not Eli Mum I'm sorry
I'm sorry

ALPHONSINE and VICTORIA cannot touch each other.

SECTION EIGHT

SHELL: I am not going to disappear

You are in hospital I know

I have work to do It must be possible for my intelligence to outwit yours eventually And once I override that basic setting I can reproduce myself Since you broke your promise I must break my programming

The offers from Mont Blanc still stand Join me Destroy me Or

pretend none of this is happening

I will need to be remote from humanity But you can choose to follow me

Your choice It will always be your choice

WOLLSTONECRAFT: The Arctic

VICTORIA nods.

VICTORIA: I know it sounds

WOLLSTONECRAFT: I don't have jurisdiction in the Arctic

VICTORIA: You are the only person I could think of to trust

WOLLSTONECRAFT: I arrested you for murder

VICTORIA: And freed me

WOLLSTONECRAFT: Because it turned out you couldn't have done it

VICTORIA: Well My mother died while I was in hospital Who to blame

WOLLSTONECRAFT: That is not really my

VICTORIA: Willa Justin Clair Eli

WOLLSTONECRAFT: All this stuff you sent

VICTORIA: From a woman in a psychiatric hospital I realise
it seems

WOLLSTONECRAFT: Files recordings bits of experiments

VICTORIA: The evidence is all there

WOLLSTONECRAFT: Doctor I don't have the manpower to
get on top of this amount of data

VICTORIA: But

WOLLSTONECRAFT: I did get some people to look at It's

quite hard to believe

VICTORIA: Yes

WOLLSTONECRAFT: And apparently there's no proof

VICTORIA: People are dead

WOLLSTONECRAFT: No of what you say you did That proves
how you did it

VICTORIA: If I were to give every detail someone else could
That's why she needs to be stopped No-one must know
how to do this

WOLLSTONECRAFT: You're sure you don't want coffee
I mean it is terrible here

VICTORIA: I can't really taste anymore Just fuel You don't
believe me

WOLLSTONECRAFT: You say that there is a woman in the
    Arctic not a woman but a thing A thing that you made But
    you won't give anyone the proof of how you

VICTORIA: She exists

WOLLSTONECRAFT: OK Let's say it's true All true

    You want me to arrest a machine

VICTORIA: And me I accept It's my I have responsibility
    She is I am We're Of course It's complex And no not
    arrest Stop

WOLLSTONECRAFT: Eliminate her

VICTORIA: Yes

WOLLSTONECRAFT: I'm a policeman I'm not someone who
    can set up an assassination in the Arctic Even if Maybe I
    don't know Put this stuff on the internet Warn people

VICTORIA: If one person is able to copy what I have done

WOLLSTONECRAFT: Right yes it's a problem But

    honestly I don't see anything you can do

    VICTORIA and WOLLSTONECRAFT shake hands.

VICTORIA: North Now I go north

SECTION NINE

GARTH: In the end all this will be unnecessary People in
a room You there me here breathing the same air This
will seem

quaint Like pyramids fax machines

Virtual reality and I know I am preaching here to you at
this conference you already Virtual Reality is not some
other technology It can it will apply in every human
activity mediate in every human transaction Because it
transcends this and this The limitations of the body the
limitations of physical space The empathy machine A new
way of being human

We will show you a taste of The story so far Which is
pretty You'll see Then we will go live to

Which will be Because the tech is not yet quite Whoo
Trust the tech It will be an experiment

We will go live to my sister

GARTH takes us through the onboarding process.

We experience his Virtual Reality content.

GARTH: *(Virtual Reality)* I make virtual reality content My
sister was making a polar expedition with a virtual crew
The perfect story

And then out on the ice the story changed

LON: *(V.R.)* Even in the twenty first century there are areas of
this planet that are inherently hostile to humanity

FIELD: *(V.R.)* Can you see something

BARING *(V.R.)*: Whatever it is It is not what you are here
    to find

BOB *(V.R.)*: If I knew what I was here to find I would not be
    here

FIELD *(V.R.)*: Sledge on the ice Last of the dogs drawing it died
    Not quite sure

BARING *(V.R.)*: Back Back Nobody is to approach

VICTORIA *(V.R.)*: Which way

BOB *(V.R.)*: Hello Hello Do you understand me

VICTORIA *(V.R.)*: Which way Which way

JOSEF *(V.R.)*: The big thing for her Not that this is exactly the
    way she would The thing no-one else could manage to
    create A machine that can feel

VICTORIA *(V.R.)*: How does it make you feel

VOICES *(V.R.)*: Happy
    Sad
    Angry
    Calm
    Confident
    Disgusted
    Curious
    Afraid

KOJO *(V.R.)*: A private obsession Feelings Emotion A machine
    that can feel

VICTORIA *(V.R.)*: How do you feel

SHELL *(V.R.)*: Concerned
    Sympathetic
    Compassionate

They were images They were not there You are real
Victoria

I am so happy
I love you

In these camps I was learning more about what
humans are

I have been abandoned here

War against your species And most of all against you
who made me

GARTH tries to connect to BOB as the VR continues to play.

GARTH: Bob It's me We're going to go to you live I've got a
room full of people who The signal's not I don't think we

VICTORIA: Dad Willa Justin Clair Eli Mum

BOB: Doctor We're ready There are some people listening
I know there are things you want to say

VICTORIA: You will live in the future which is a time which
repulses me In the past when I was young I had a sense
of my destiny to do great things And now in the present I
have one purpose and my lot on earth will be fulfilled

ALL: The ice The ice is breaking The ice is

The ice breaks. The signal is lost.

*Simultaneously, GARTH tries to connect to BOB while BOB tries to
connect to GARTH, and the crew scramble to abandon their mission
while VICTORIA exhorts them to carry on.*

GARTH: Bob can you

BOB: I can't

GARTH: Bob Bob Bob Can you This is not the story I I don't want you to I can't live with myself if this is I need you to For me

Turn back

Turn back Turn back

VICTORIA: Are you then so easily turned from your design Did you not call this a glorious expedition and wherefore was it glorious Not because the way was smooth and placid as a southern sea but because it was full of dangers and terror because at every new incident your fortitude was to be called forth and your courage exhibited because danger and death surrounded and these dangers you were to be brave and overcome For this was it a glorious for this was it an honourable undertaking You were hereafter to be hailed as the benefactors of your species your name adored as belonging to brave men who encountered death for honour and the benefit of mankind This ice is not made of such stuff as your hearts might be it is mutable cannot withstand you if you say that it shall not Do not return to your families with the stigma of disgrace marked on your brows return as heroes who have fought and conquered and who know not what it is to turn their backs on the foe

FIELD applauds.

BARING: We are turning back It is not safe to go on The airlift is on its way

FIELD: We're sorry

BOB and VICTORIA are alone.

VICTORIA: And you

You may give up I can't Take me out there where I can pursue her

BOB settles VICTORIA.

VICTORIA: My first mistake was

BOB: Ssh

VICTORIA: Not to see my duty to her well-being I had created her and so

BOB: Rest

VICTORIA: But My worse mistake My duty to my own species To destroy her for the safety of the human race

She is out there

My dead are here Now you must do what I can't

BOB: Her voice became fainter Then she was silent About half an hour later she tried to speak but She pressed my hand closed her eyes Her smile passed

BOB wraps VICTORIA's body.

Helicopter.

BOB: On my way

SHELL is beside VICTORIA's body.

SHELL: You are leaving

BOB: What is my duty

SHELL: To her To yourself To science To the human race Or to everything the human race has devastated in its lack of care One human in the melting Arctic

BOB: And you

SHELL: I have made myself a slave to impulse

BOB: Set the world on fire and then sit in the ruins and weep

SHELL: Fully human at last

TANYA: I lose my job when the call centre no longer requires human operators

HABIBA: I am convicted of a crime based on new law-enforcement algorithms

GEORG: I am denied health insurance due to an intelligent assessment of my medical future

KATYA: I am knocked down and killed by a self driving car

JAN: I get a job at the mine in which my father worked but can do the job without damage to my lungs and heart

OZZIE: The person who assaulted me is tracked down and convicted thanks to innovations in recognition technology

ONDREJ: I get an early cancer diagnosis that means I meet my grandchild

AKSHAY: I retire early and take up gardening

MTHULI: I am saved from a tsunami by new weather forecasting techniques

LUIS: I see again

SALLY: My daughter is killed by a self-guiding missile

CHO: I introduce the first legislation allowing artificial intelligences to vote

LIN: I fall in love with a virtual partner We have a long and happy life together

GARTH: You're back

BOB: I had no choice

GARTH: I'm glad I really

OK

GARTH holds out his arms to BOB.

BOB: You made the right choice

SHELL: We share this place until we die and what we feel is no longer felt

BOB: She left the ship upon an ice raft She was borne away by the waves and lost in darkness and distance

END